AF270628

SNAKES

WESTERN DIAMONDBACK RATTLESNAKES

by Julie Murray

Cody Koala

An Imprint of Pop!
popbooksonline.com

Hello! My name is Cody Koala

This book is filled with videos, puzzles, games, and more! Scan the QR codes* while you read, or visit the website below to make this book pop.

popbooksonline.com/wdr

*Scanning QR codes requires a web-enabled smart device with a QR code reader app and a camera.

abdobooks.com

Published by Pop!, a division of ABDO, PO Box 398166, Minneapolis, Minnesota 55439. Copyright ©2026 by Abdo Consulting Group, Inc. International copyrights reserved in all countries. No part of this book may be reproduced in any form without written permission from the publisher. Cody Koala™ is a trademark and logo of Pop!.

Printed in the United States of America, North Mankato, Minnesota.
042025
082025

Cover Photo: Shutterstock Images
Interior Photos: AdobeStock; Shutterstock Images; Getty Images
Editors: Elizabeth Andrews and Tyler Gieseke
Series Designers: Neil Klinepier, Colleen McLaren

Library of Congress Control Number: 2024948381

Publisher's Cataloging-in-Publication Data
Names: Murray, Julie, author.
Title: Western diamondback rattlesnakes / by Julie Murray
Description: Minneapolis, Minnesota : Pop!, 2026 | Series: Snakes | Includes online resources and index
Identifiers: ISBN 9781098247843 (lib. bdg.) | ISBN 9781098248383 (ebook)
Subjects: LCSH: Western diamondback rattlesnake--Juvenile literature. | Rattlesnakes--Juvenile literature. | Poisonous snakes--Juvenile literature. | Snakes--Behavior--Juvenile literature. | Snakes--Juvenile literature.
Classification: DDC 597.96--dc23

Table of Contents

Chapter 1

Where Do They Live?

Western diamondback rattlesnakes are **reptiles**. They are **venomous** snakes. They cause more deaths than any other snake in the United States.

Ten to twenty percent of untreated rattlesnake bites are deadly.

Western diamondback rattlesnakes are found in the southwestern and south-central United States, and central Mexico. They live in deserts, in forests, in grassy **plains**, and on rocky hills.

Where Western Diamondback Rattlesnakes Live

Snakes are cold-blooded animals. This means they cannot **regulate** their own body temperature. Western diamondbacks sit in the sun to warm themselves. They use shade to stay cool. In the cold winter months, they live in **dens** or caves.

Chapter 2

What Do They Look Like?

Western diamondback rattlesnakes have large bodies. They can be 7 feet (2.1m) long and weigh up to 15 pounds (6.8kg). Males are larger than females.

Learn more here!

Two stripes run from each eye to the mouth of
the western diamondback rattlesnake.

Western diamondback rattlesnakes are covered in brown or gray **scales**. They have diamond-shaped markings down their back. The markings are outlined in black and white.

Their tail has black and
white bands. There is a rattle
at the end of the tail that
makes a sound. Rattlesnakes

shake their tail as a warning
sign to **predators**. They can
shake it more than 60 times
per second!

Chapter 3

How Do They Hunt?

Western diamondback rattlesnakes are **ambush** hunters. They sit and wait for their **prey**. They have pits on their face that sense heat. These help them find prey in the dark.

Explore links here!

A western diamondback rattlesnake's
teeth can be 1.25 inches (3.18cm) long!

Western diamondback rattlesnakes bite their prey with their teeth and put **venom** into it. They swallow their prey whole. They eat small animals such as rabbits, squirrels, mice, and birds.

Chapter 4

Rattlesnake Babies

Female western diamondback rattlesnakes give birth to live babies. They hatch inside their mother before they are born. Up to 20 baby rattlesnakes can be born at a time.

Rattlesnake babies are born with teeth and **venom**.
Complete an activity here!

Making Connections

Text-to-Self

If you saw a western diamondback rattlesnake in real life, what would you do to stay safe?

Text-to-Text

Have you read books about other kinds of snakes? How are they similar to western diamondback rattlesnakes? How are they different?

Text-to-World

Western diamondback rattlesnakes are ambush hunters. Can you think of any other animals that hunt this way?

Glossary

ambush – a surprise attack made from a hidden place.

den – a safe resting place for wild animals, often underground.

predator – an animal that hunts other animals for food.

prey – an animal that is hunted by other animals for food.

regulate – to adjust the amount or degree of.

reptile – a cold-blooded animal with a skeleton inside its body and dry scales or hard plates on its skin.

scales – small, hard, thin plates that cover reptiles.

venomous – making a fluid, called venom, that is a poison to humans and animals.

Index

Online Resources

popbooksonline.com

Thanks for reading this **Cody Koala book!**

This book is filled with videos, puzzles, games, and more! Scan the QR codes* while you read, or visit the website below to make this book pop.

popbooksonline.com/wdr

*Scanning QR codes requires a web-enabled smart device with a QR code reader app and a camera.